Apollo 11: 52nd Anniversary Pictorial

By Donald Yates

Independently Published by Teamyates Enterprises LLC

ISBN: 9798540600828

Copyright © 2021 Donald Yates All rights reserved.

No part of this book may be reproduced, or stored in a retrieval system, or transmitted in any form or by any means, electronic, mechanical, photocopying, recording, or otherwise, without express written permission of the publisher.

Contents

Introduction/Dedication 1
Saturn 5 Rocket Launch 5
Saturn 5 Base Firing Closeups 9
Mission Control 14
Trans-Lunar Phase 17
Separation of Lunar Module/Command Post 31
At the Moon 36
Recovery 82
Appendix A Star Chart 85
Appendix B Flight Profile 87
Appendix C Flight Plan 88
References 89
Other Books by the Author 90

Donald Yates

Introduction/Dedication

Thank you for purchasing *Apollo 11: 52nd Anniversary Pictorial*.

What do you think was the most popular reality show in history? Maybe the Muhammad Ali-Joe Frazier fight or a show about who gets to marry a handsome bachelor or beautiful bachelorette. Those two events do not come close to the over 600 million people who watched the Apollo 11 moon landing, more specifically Neil Armstrong taking the first step on the moon by a human. The Apollo 11 mission was seen in virtually every nation world-wide and was the most viewed event ever up to that time. Viewers stayed glued to their televisions for hours and hours watching this unheard of event as it happened.

I can remember watching the Apollo 11 mission myself as a six-year-old child. I normally couldn't sit still for more than a minute or two but I sat and watched in amazement for what seemed like hours. I surprised myself sitting in one place for so long for something other than cartoons. I can still recall watching the Apollo 11 mission as if it happened a week or two ago.

Apollo 11: 52nd Anniversary Pictorial

The Apollo 11 mission was medicine for America. In the late 1960s, our nation was in turmoil. There were riots and assassinations occurring regularly. Our nation was embroiled in an ugly war in Vietnam that we weren't allowed to win. There was also the seemingly never ending Cold War tension between the United States and the Soviet Union.

Apollo 11's success gave America a much needed break from all the bad news. It reminded us that despite all our challenges and disappointments, we were still the world's greatest nation.

It has now been over 50 years since that memorable first step on the moon. It has been over a decade since Neil Armstrong passed away. You are probably wondering why I'd add this book to a pool of many fine Apollo 11 books. This event was so pivotal to world history. You can never learn too much about the Apollo 11. I've learned more by putting together this book and hope you will too.

Most of the photos you are about to see were taken by the astronauts themselves. A few, especially the photos of the actual launching, were screenshots taken from very good NASA video

Donald Yates

footage. These photos were obtained from the National Archives and Records Administration (NARA) in Washington D.C. Keep in mind that the this book's author and publisher are in no way affiliated with NASA or the United States government. I want to thank the staff at NARA for making these images available. I also appreciate the incredible job done by NASA in recording the events of the Apollo 11 mission back in 1969. These photos and videos will be priceless treasures for all the world for the rest of humankind's existence.

I dedicate this book on the 52nd Anniversary of the Apollo 11 mission launch to the astronauts of all the Apollo missions.

Apollo 11: 52nd Anniversary Pictorial

The Apollo 11 crew, Neil A. Armstrong, Michael Collins, and Edwin "Buzz" Aldrin.

Donald Yates

Saturn 5 Rocket Launch

The Apollo 11 Saturn V SA-506 rocket just before takeoff.

Apollo 11: 52nd Anniversary Pictorial

Saturn V SA-506 liftoff burns just after firing.

Saturn V SA-506 lifts off. Destination is the moon.

Apollo 11: 52nd Anniversary Pictorial

Donald Yates

Saturn 5 Base Firing Closeups

This is the first of five photos showing the base of the Saturn 5 rocket during firing. The above photo was taken just before liftoff. Notice the two tower like frame devices on the right side of the Saturn 5 rocket. These devices have arms that support the rocket until precise time to lift off. There is actually a third support on the other side of the rocket that is only partially visible in this photo. See the references section in this book to get the web address to watch the close up, 500 frames a second slow motion video of the base of the rocket during lift off.

Apollo 11: 52nd Anniversary Pictorial

Base of the Saturn 5 rocket just after initiation of firing sequence. The support arms are still stabilizing the rocket.

The rocket firing becomes more intense and incredibly hot. The arms still hold though. The debris falling at the top of the photo is actually ice.

Apollo 11: 52nd Anniversary Pictorial

Notice that the supporting arms have now released the rocket. The arms are moving up into the frame housings to keep them safely away from the rocket.

The support arms have swung up into those frame towers. The rocket is now beginning to gain altitude.

Apollo 11: 52nd Anniversary Pictorial

Mission Control

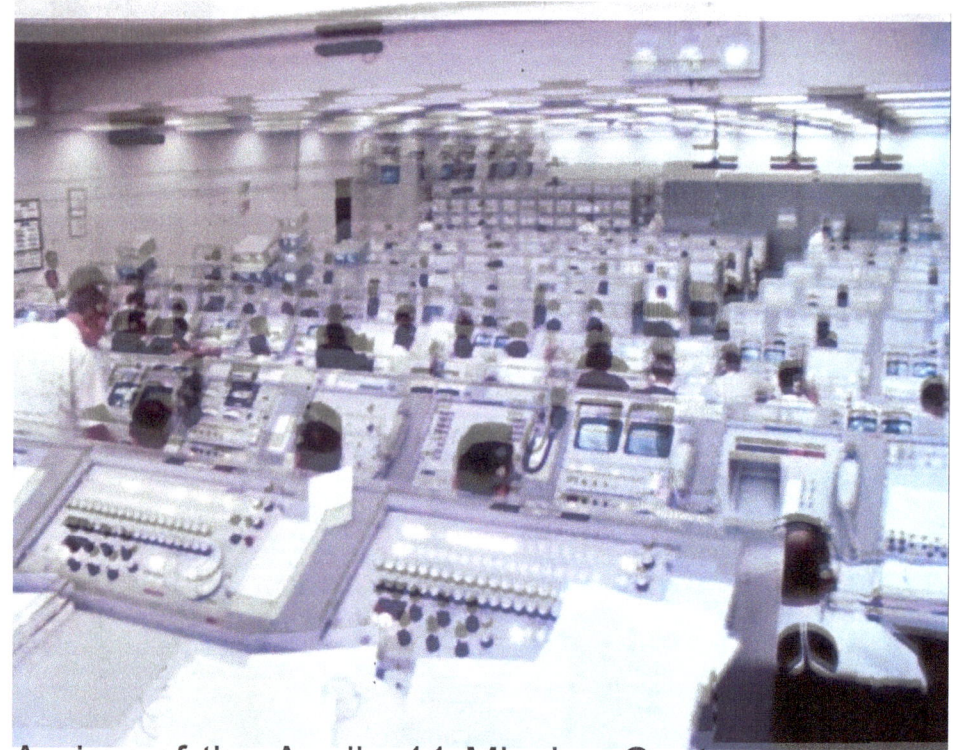

A view of the Apollo 11 Mission Control Center in Houston, Texas.

Mission Control Center staff watch closely as the Saturn V SA-506 rocket travels towards the heavens.

Apollo 11: 52nd Anniversary Pictorial

Mission Control Center staff and visitors applaud a successful liftoff.

Donald Yates

Trans-Lunar Phase

Spacecraft interior. Command and Service Module. Image taken during translunar phase of the Apollo 11 mission.

Apollo 11: 52nd Anniversary Pictorial

Earth limb with clouds. Glare of Sun visible above the Earth's horizon. Image taken during translunar phase of the Apollo 11 mission.

Earth limb with view of Tropical Storm Bernice in the Pacific Ocean off the coast of Baja California. Image taken during translunar phase of the Apollo 11 mission.

Apollo 11: 52nd Anniversary Pictorial

Earth view over Central and North America. Image taken during the translunar phase of the Apollo 11 mission.

An interior view of the Apollo 11 Lunar Module showing displays and controls during checkout and inspection after hatch opening. Mounted in the Lunar Module window is a 16mm data acquisition camera which has a variable frame speed of 1, 6, 12 and 24 frames per second. Image taken during the translunar phase of the Apollo 11 mission.

Apollo 11: 52nd Anniversary Pictorial

Earth view with clouds. Image taken during the translunar phase of the Apollo 11 mission.

A close-up view of the docking target on the Apollo 11 Lunar Module photographed from the Command and Service Module during the Lunar Module/Command and Service Module docking in lunar orbit. This image was taken during the translunar phase of the Apollo 11 mission.

Apollo 11: 52nd Anniversary Pictorial

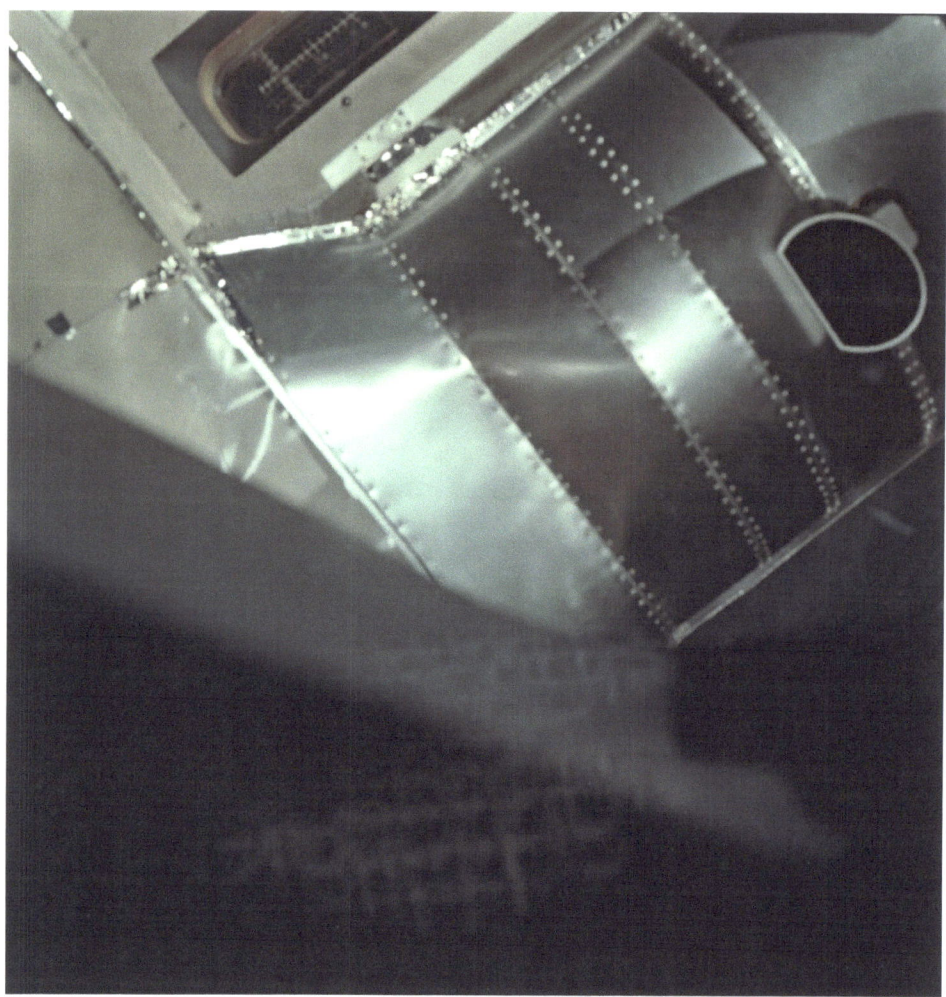

Close up of Lunar Module attached to the Saturn Launch Vehicle Third Stage (SIVB) as the Command and Service Module (CM) docks with the Lunar Module. Image taken from the CM during the translunar phase of the Apollo 11 mission.

Saturn Launch Vehicle Third Stage (S-IVB) with Lunar Module attached. Image taken from the Command and Service Module as it moves into dock with the Lunar Module after the adapter panels have been ejected. Image taken during the translunar phase of the Apollo 11 mission.

Apollo 11: 52nd Anniversary Pictorial

Astronaut Edwin E. Aldrin Jr., Lunar Module pilot, inside the Lunar Module during checkout and inspection after hatch opening. Image taken during the translunar phase of the Apollo 11 mission by Astronaut Neil A. Armstrong, Commander of the Apollo 11 mission.

An Apollo 11 astronaut sets up the onboard television camera while inside the Lunar Module tunnel after Lunar Module hatch opening. Image taken during the translunar phase of the Apollo 11 mission.

Apollo 11: 52nd Anniversary Pictorial

Image taken during the translunar phase of the Apollo 11 mission.

A close-up view of the docking target on the Apollo 11 Lunar Module photographed from the Command and Service Module during the Lunar Module/Command and Service Module docking in lunar orbit. This image was taken during the translunar phase of the Apollo 11 mission.

Apollo 11: 52nd Anniversary Pictorial

Earth view over Central and North America. Image taken during the translunar phase of the Apollo 11 mission.

Donald Yates

Separation of Lunar Module/Command Post

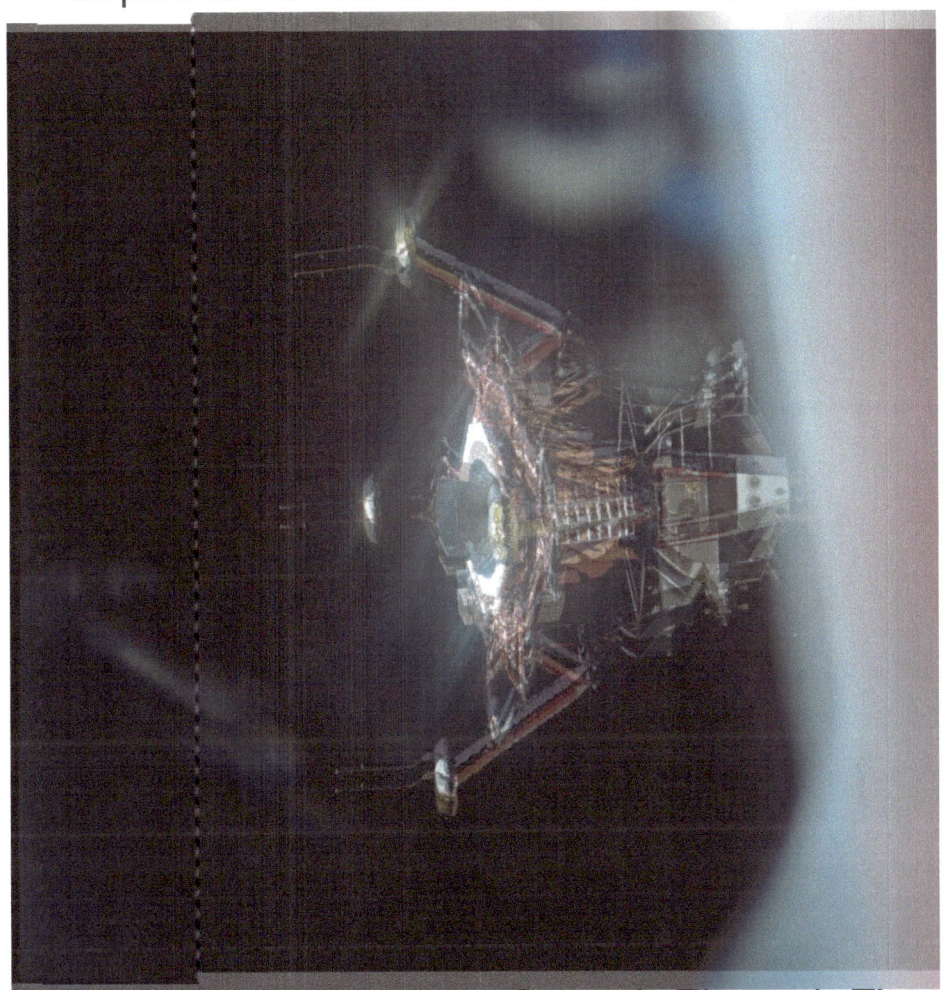

View of Lunar Module; Ground Elapsed Time (GET): 100:30. This image was taken during separation of the Lunar Module and the Command Module during Apollo 11 mission.

Apollo 11: 52nd Anniversary Pictorial

View of Lunar Module. This image was taken during separation of the Lunar Module and the Command Module during Apollo 11 mission. Blackness of space in background.

View of Lunar Module. This image was taken during separation of the Lunar Module and the Command Module during the Apollo 11 mission. Blackness of space in background.

View of Lunar Module and footpads, Ground Elapsed Time (GET): 100:30. This image was taken during separation of the Lunar Module and the Command Module during Apollo 11 mission.

View of Lunar Module. This image was taken during separation of the Lunar Module and the Command Module during Apollo 11 mission. Blackness of space in background.

Apollo 11: 52nd Anniversary Pictorial

At the Moon

View of the Moon limb, Lunar Module during ascent, Mare Smythii. Earth begins to appear on the horizon. Image was taken during the Apollo 11 mission.

View of the Moon limb, Lunar Module during ascent, Mare Smythii. Earth appears on the horizon. Image was taken during the Apollo 11 mission.

Apollo 11: 52ⁿᵈ Anniversary Pictorial

This grainy photo from a video screen shot shows astronaut Neil Armstrong (at right) shortly after taking the first step on the moon by a human.

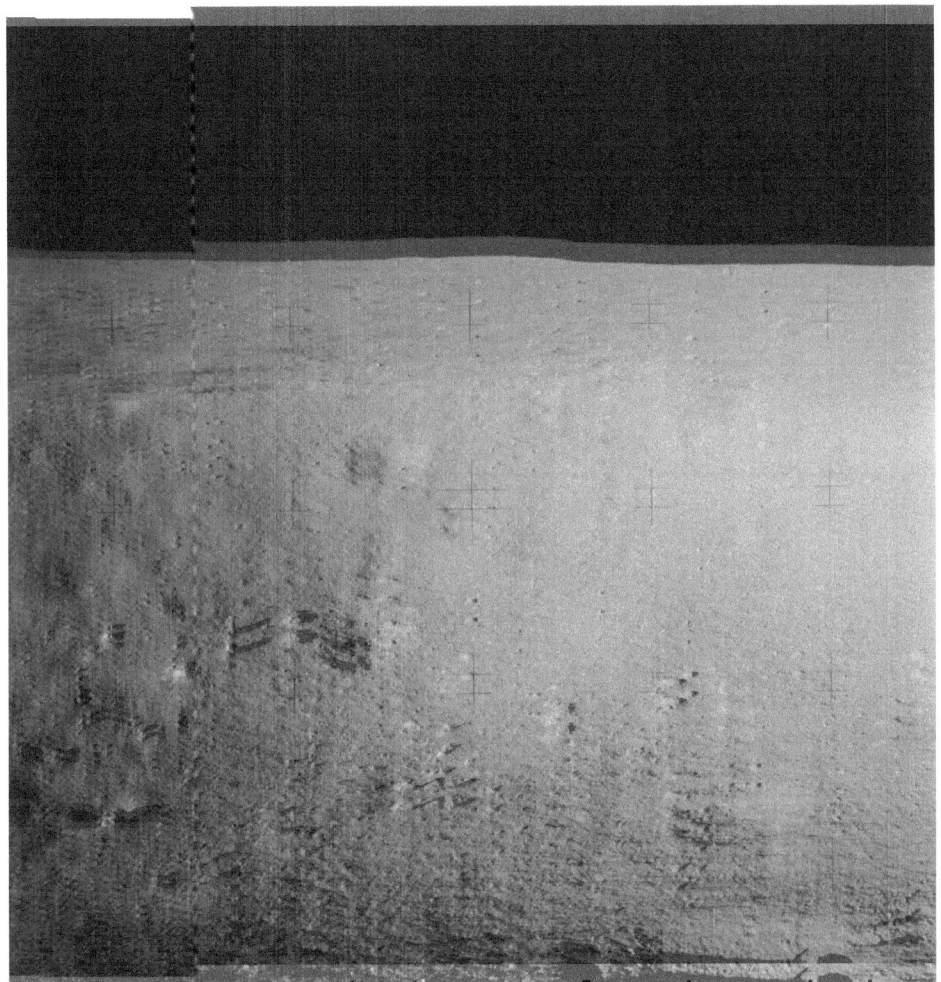

Small craters on the lunar surface. Lunar horizon visible. This image is the first of a multi-framed panorama photographed from a point some 30 or 40 feet west of the plus-Z (west) footpad of the Lunar Module "Eagle." The view is looking toward the southwest showing part of the horizon crater rim that was pointed out as being visible from the Eagle's window. Image taken at Tranquility Base during the Apollo 11 mission.

Apollo 11: 52nd Anniversary Pictorial

Oblique towards Tranquility Base, part coverage of Target of Opportunity (TO) 115. TO 115 is Craters Hypatia and Alfraganus A, and chain of shallow sharp depressions East of Alfraganus. Image taken from inside the Lunar Module from orbital altitude over the moon during the Apollo 11 mission.

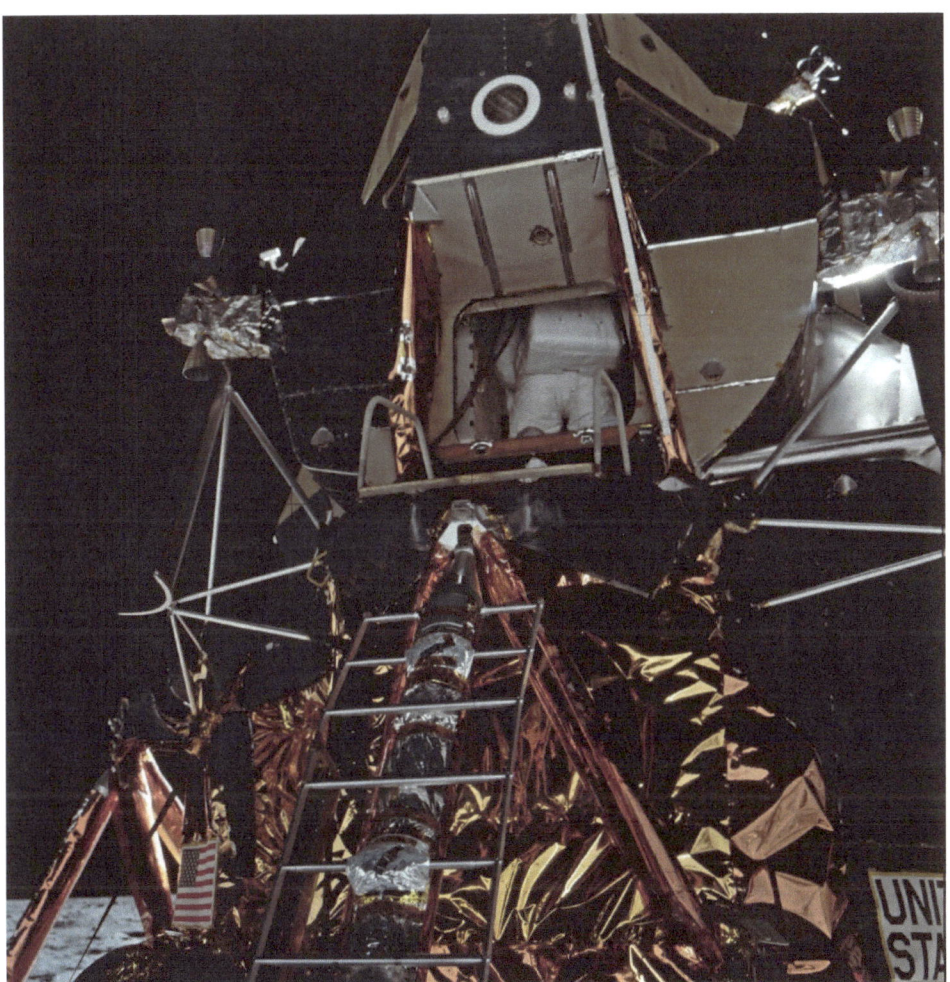

View looking up the Lunar Module ladder as Astronaut Edwin E. Aldrin Jr., the Lunar Module pilot, prepares to egress the module. Image taken at Tranquility Base during the Apollo 11 mission.

Apollo 11: 52nd Anniversary Pictorial

Astronaut Edwin E. Aldrin Jr., the Lunar Module pilot, egresses the Lunar Module. Image taken at Tranquility Base during the Apollo 11 mission.

Astronaut Edwin E Aldrin Jr, Lunar Module pilot descends from the Lunar Module, climbing down the ladder. Lunar horizon visible in background. Image taken at Tranquility Base during the Apollo 11 mission.

Apollo 11: 52nd Anniversary Pictorial

Astronaut Edwin E Aldrin Jr, Lunar Module pilot descends from the Lunar Module, climbing down the ladder. Lunar horizon visible in background. Image taken at Tranquility Base during the Apollo 11 mission.

Astronaut Edwin E Aldrin Jr, Lunar Module pilot, climbs down the Lunar Module ladder, preparing for his first steps on the moon. Lunar horizon visible in background. Image taken at Tranquility Base during the Apollo 11 mission.

Apollo 11: 52nd Anniversary Pictorial

Close up view of Lunar Module skirt: Image taken at Tranquility Base during the Apollo 11 mission.

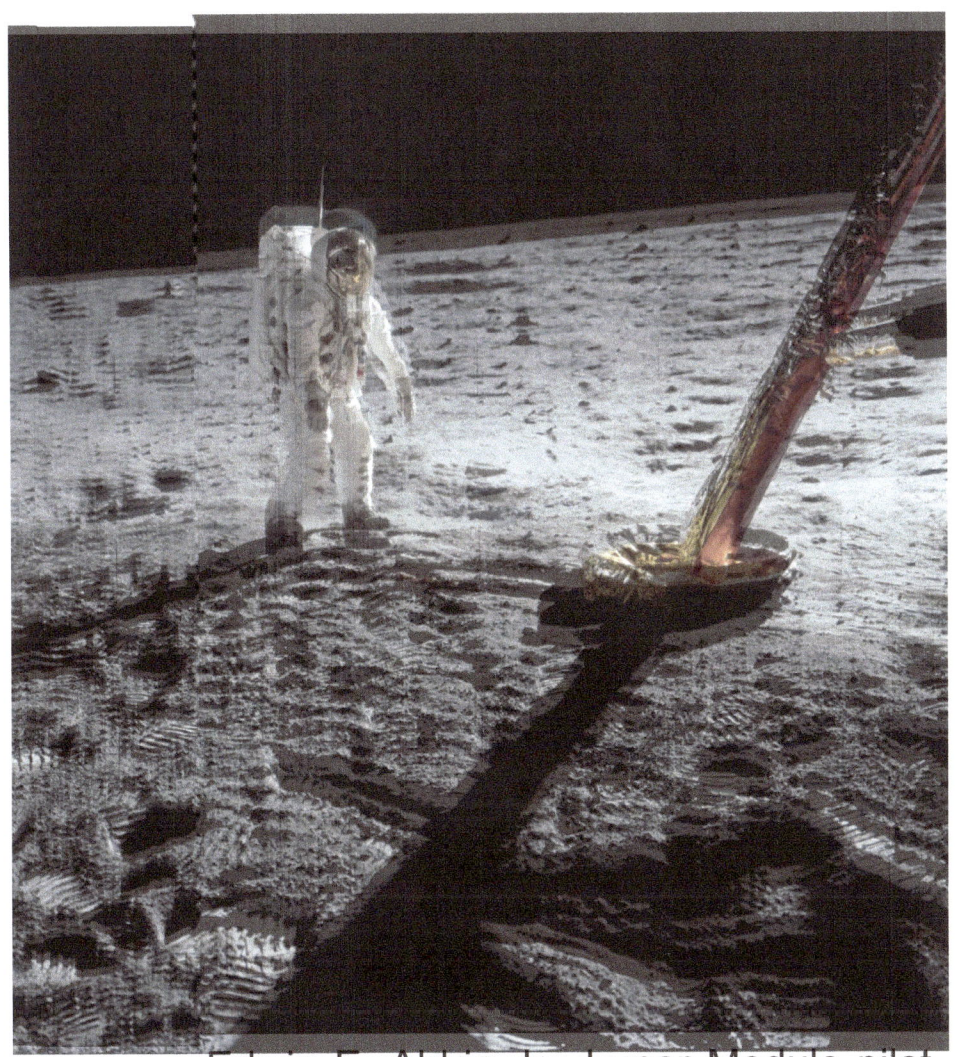

Astronaut Edwin E. Aldrin Jr., Lunar Module pilot, walks near the Lunar Module. Lunar Module strut and footpad visible. Astronaut boot prints are visible on the lunar surface. Image taken at Tranquility Base.

Apollo 11: 52nd Anniversary Pictorial

Astronaut Edwin E. Aldrin, Lunar Module pilot, salutes the U.S. flag that has been placed on the moon. The Lunar Module is visible in the left field of view. Numerous footprints and the cable of the surface television camera are visible on the lunar surface in the foreground. Image taken at Tranquility Base during the Apollo 11 mission.

Astronaut boot print on the lunar surface. Image taken at Tranquility Base during the Apollo 11 mission.

Apollo 11: 52nd Anniversary Pictorial

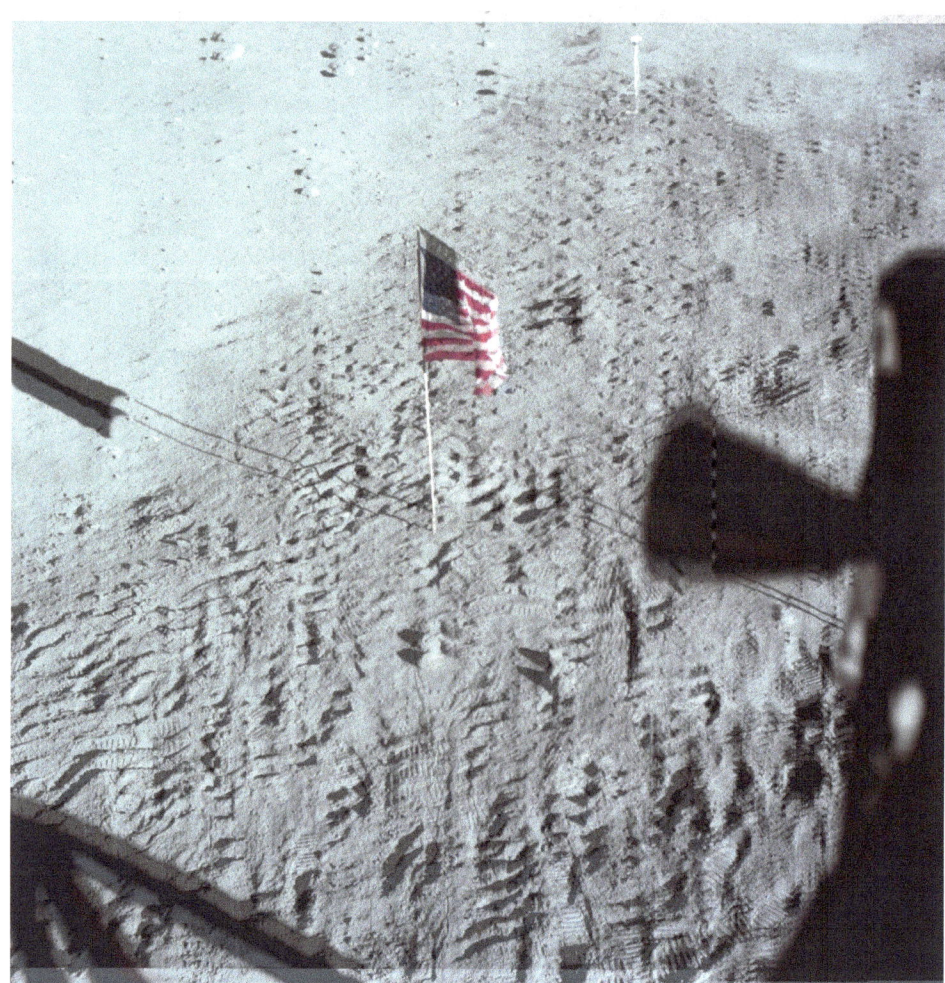

Lunar surface at Tranquility Base; the Lunar Module landing site. Lunar Module thruster shadow, United States flag and television camera are visible on lunar surface. Image taken from inside the Lunar Module during the Apollo 11 mission.

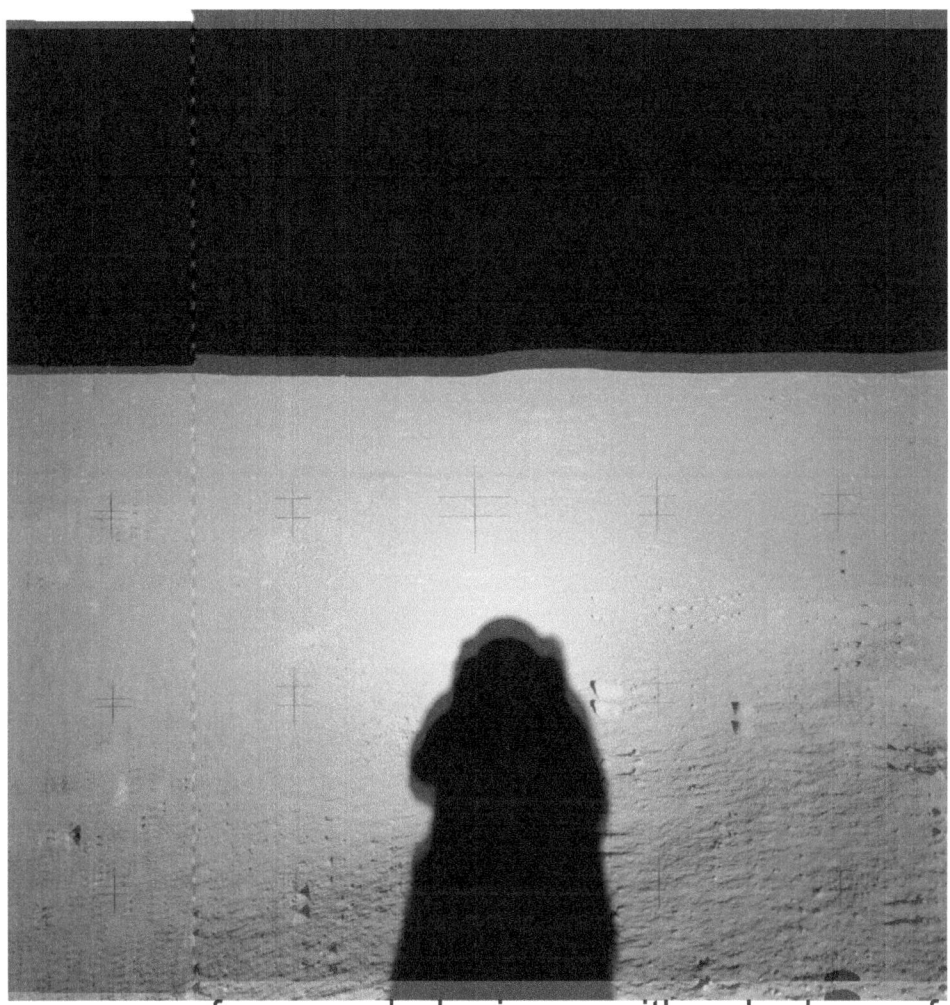

Lunar surface and horizon with shadow of astronaut in foreground as he takes the photo. Image taken at Tranquility Base during the Apollo 11 mission.

Apollo 11: 52nd Anniversary Pictorial

Lunar surface with an astronaut boot and boot print in field of view. Image taken at Tranquility Base during the Apollo 11 mission.

Astronaut Edwin E. Aldrin, Jr., Lunar Module pilot, unpacks the Early Apollo Scientific Experiments Package (EASEP) from the Modularized Equipment Storage Assembly (MESA) of the Lunar Module. Image taken at Tranquility Base during the Apollo 11 mission.

Apollo 11: 52nd Anniversary Pictorial

Astronaut Edwin E. Aldrin, Lunar Module pilot, takes a core-tube sample. The Solar-Wind Composition (SWC) experiment is visible in front of Aldrin. Image taken at Tranquility Base during the Apollo 11 mission.

Astronaut Edwin E. Aldrin, Lunar Module pilot, deploys the Passive Seismic Experiments Package (PSEP). The Laser Ranging Retroreflector (LRRR) is visible behind him to the left. The Lunar Module is in the background. The television camera and the Apollo Lunar Surface Closeup Camera (ALSCC) are visible also. Image taken at Tranquility Base during the Apollo 11 mission.

Apollo 11: 52nd Anniversary Pictorial

Spacecraft interior. Astronaut Neil A. Armstrong, Apollo 11 commander, looking at the camera from inside the Command and Service Module. Image is out of focus. Image taken during translunar phase of the Apollo 11 mission.

Astronaut Edwin E. Aldrin, Lunar Module pilot, sets up the Solar-Wind Composition (SWC) Experiment, facing it toward the sun. The Lunar Module sits to the right and slightly behind Aldrin. Linear trails (lines) in the foreground going towards the Lunar Module were formed by the cable of the surface television camera. The cable is partially visible. Image taken at Tranquility Base during the Apollo 11 mission.

Apollo 11: 52nd Anniversary Pictorial

President Nixon speaks to the astronauts by phone from the White House.

This grainy video screenshot was taken during President Nixon's phone call (previous photo) to the astronauts. The astronaut on the right is saluting toward the camera that is displaying these activities on television to the President as well as the rest of the world. The astronaut on the left has already lowered his salute.

Apollo 11: 52nd Anniversary Pictorial

Lunar surface and horizon with U.S. Flag visible. Shadow in foreground. Image taken at Tranquility Base during the Apollo 11 mission.

Z Lunar Module footpad in foreground with the +Y footpad in background. Image taken at Tranquility Base during the Apollo 11 mission.

Apollo 11: 52nd Anniversary Pictorial

Astronaut Edwin E. Aldrin Jr., Lunar Module pilot, walks near the module as a picture is taken of him. Discoloration is visible on his boots and suit from the lunar soil adhering to them. Reflection of the Lunar Module and Astronaut Neil A. Armstrong is visible in Aldrin's helmet visor. Image taken at Tranquility Base during the Apollo 11 mission.

Astronaut Edwin E. Aldrin, Lunar Module pilot, stands beside the Passive Seismic Experiments Package (PSEP). The Laser Ranging Retroreflector (LRRR), U.S. Flag, television camera and the Apollo Lunar Surface Closeup Camera (ALSCC) and Lunar Module are visible also. Image taken at Tranquility Base during the Apollo 11 mission.

Apollo 11: 52nd Anniversary Pictorial

Apollo 11 mission image - far side terminator.

Earth limb. Image taken during translunar phase of the Apollo 11 mission.

Apollo 11: 52nd Anniversary Pictorial

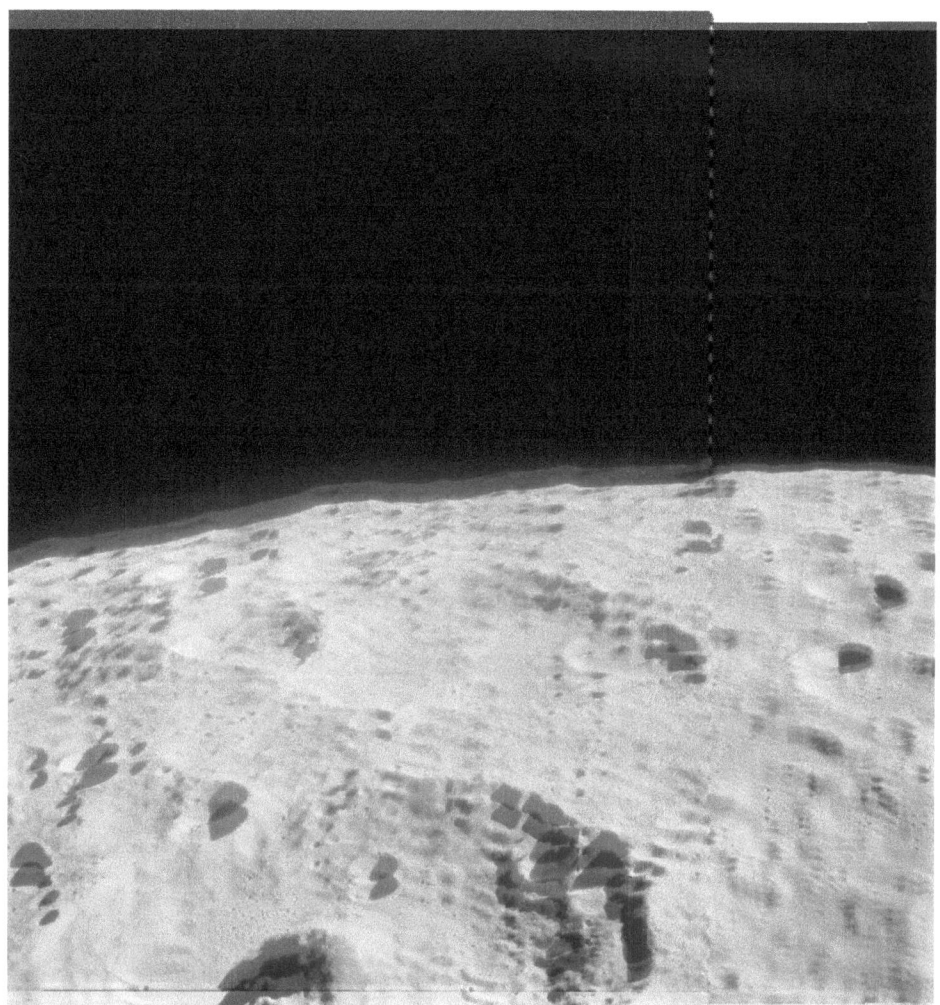

View of Moon; Crater 310, officially named Icarus. Image taken from the Command and Service Module during the Apollo 11 mission while on a near circular lunar equatorial orbit.

View of Moon, Maskelyne G. Image taken from the Command and Service Module during the Apollo 11 mission while on a near circular lunar equatorial orbit.

Apollo 11: 52nd Anniversary Pictorial

View of Moon, Secchi K. Image taken from the Command and Service Module during the Apollo 11 mission while on a near circular lunar equatorial orbit.

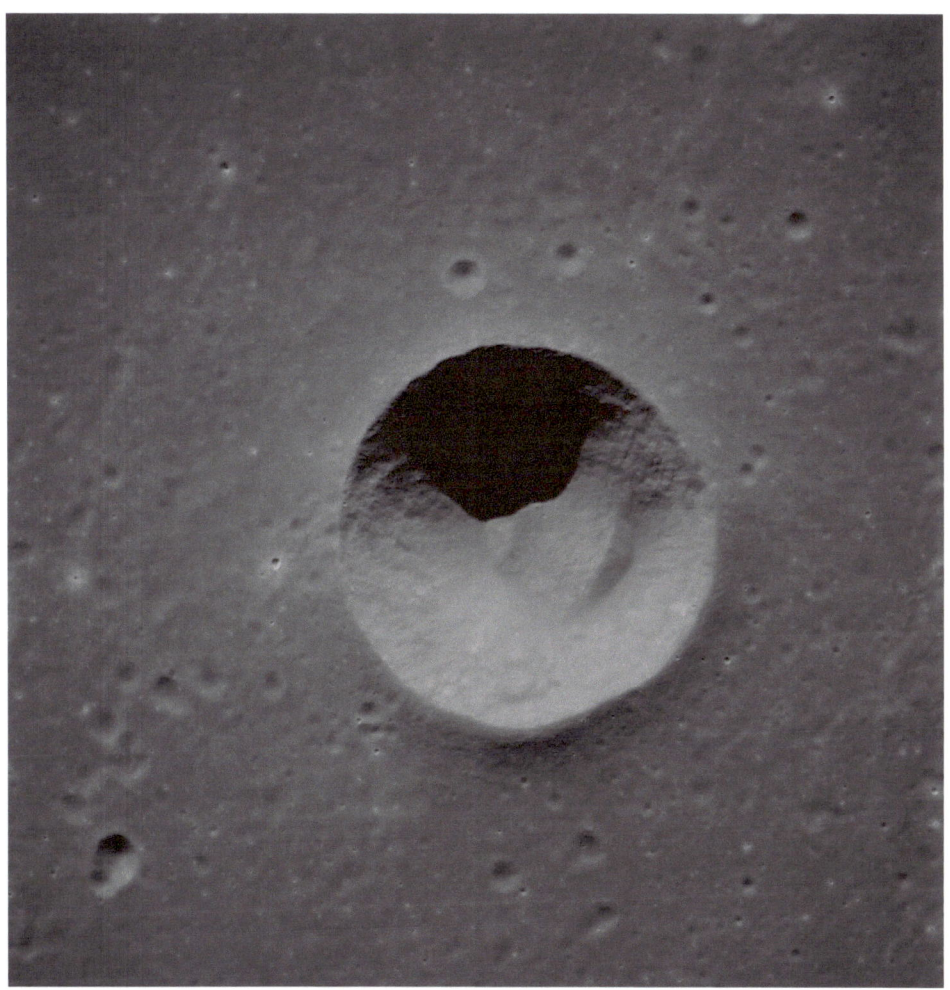

View of Moon, Taruntius G. Image taken from the Command and Service Module during the Apollo 11 mission while on a near circular lunar equatorial orbit.

Apollo 11: 52nd Anniversary Pictorial

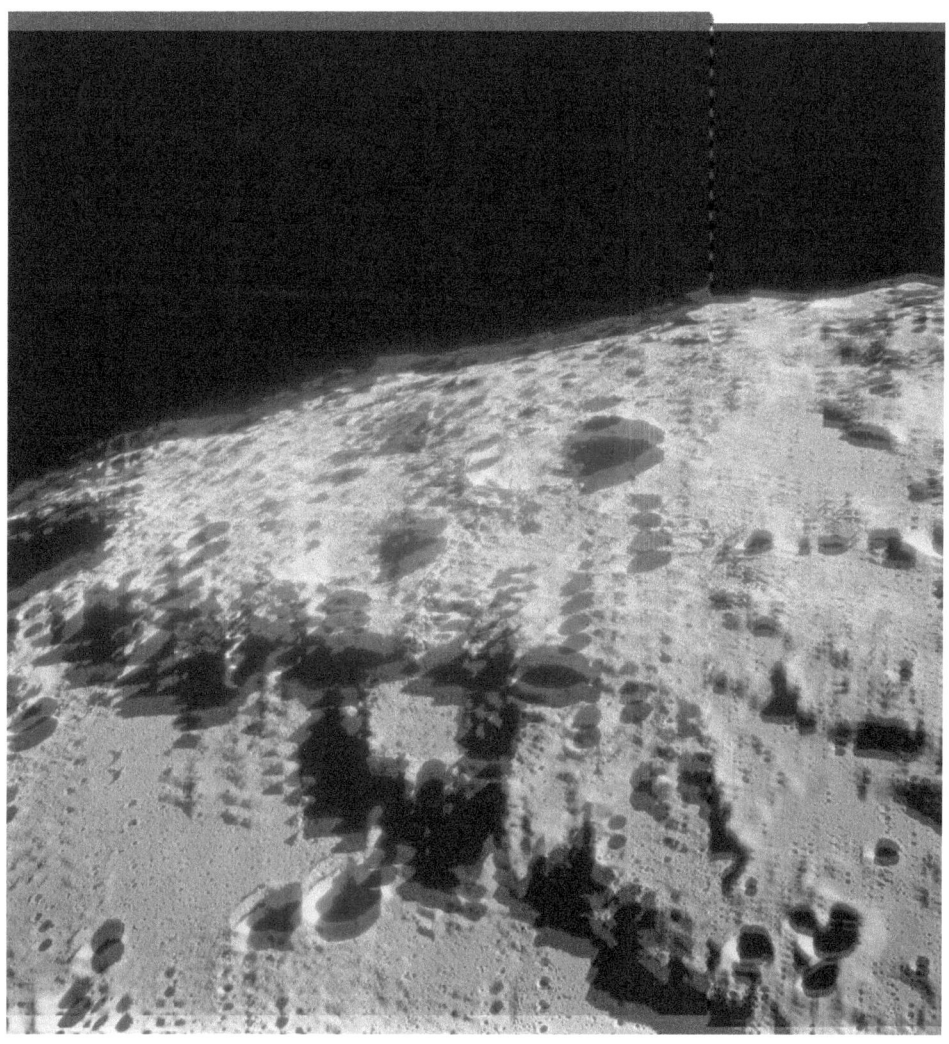

View of Moon; Crater 312; officially named Crookes. Image taken from the Command and Service Module during the Apollo 11 mission while on a near circular lunar equatorial orbit.

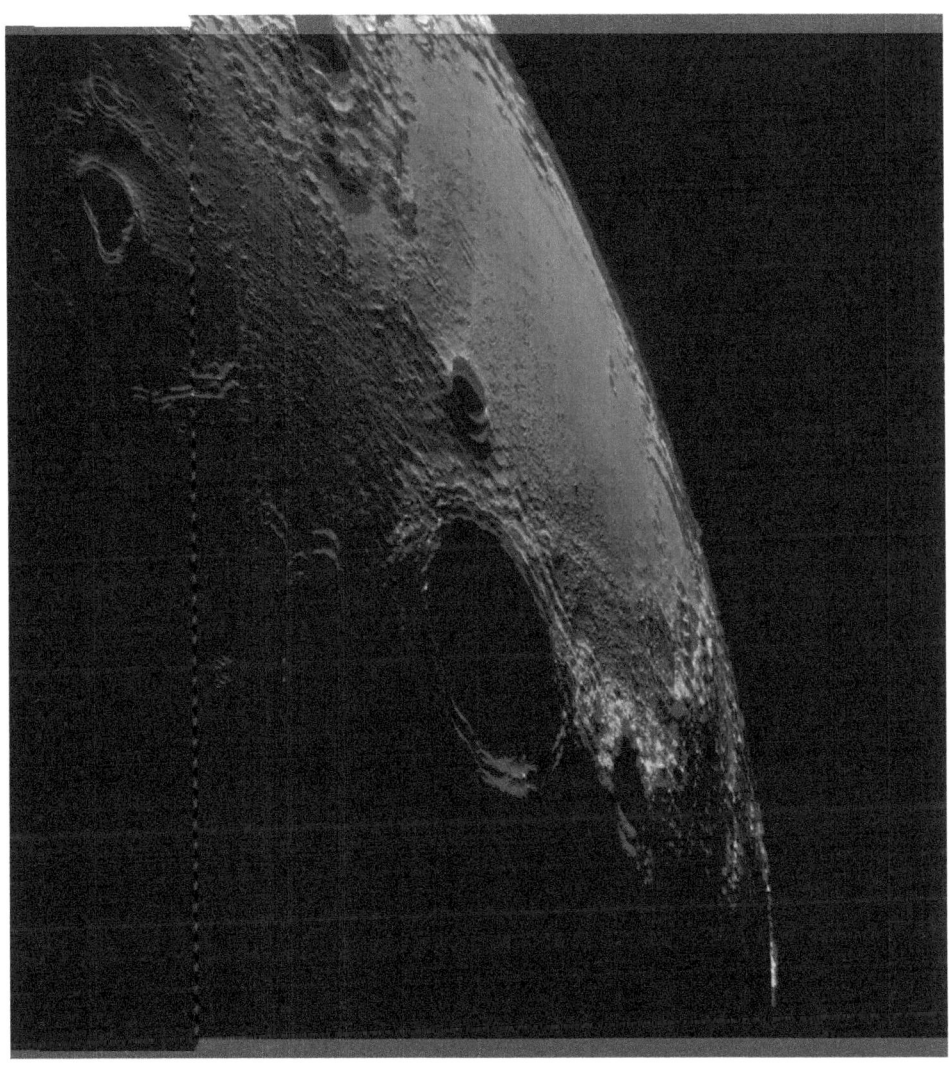

View of Moon; Theophilus: Image taken from the Command and Service Module during the Apollo 11 mission while on a near circular lunar equatorial orbit:

Apollo 11: 52nd Anniversary Pictorial

View of Moon, Maskelyne. Image taken from the Command and Service Module during the Apollo 11 mission while on a near circular lunar equatorial orbit.

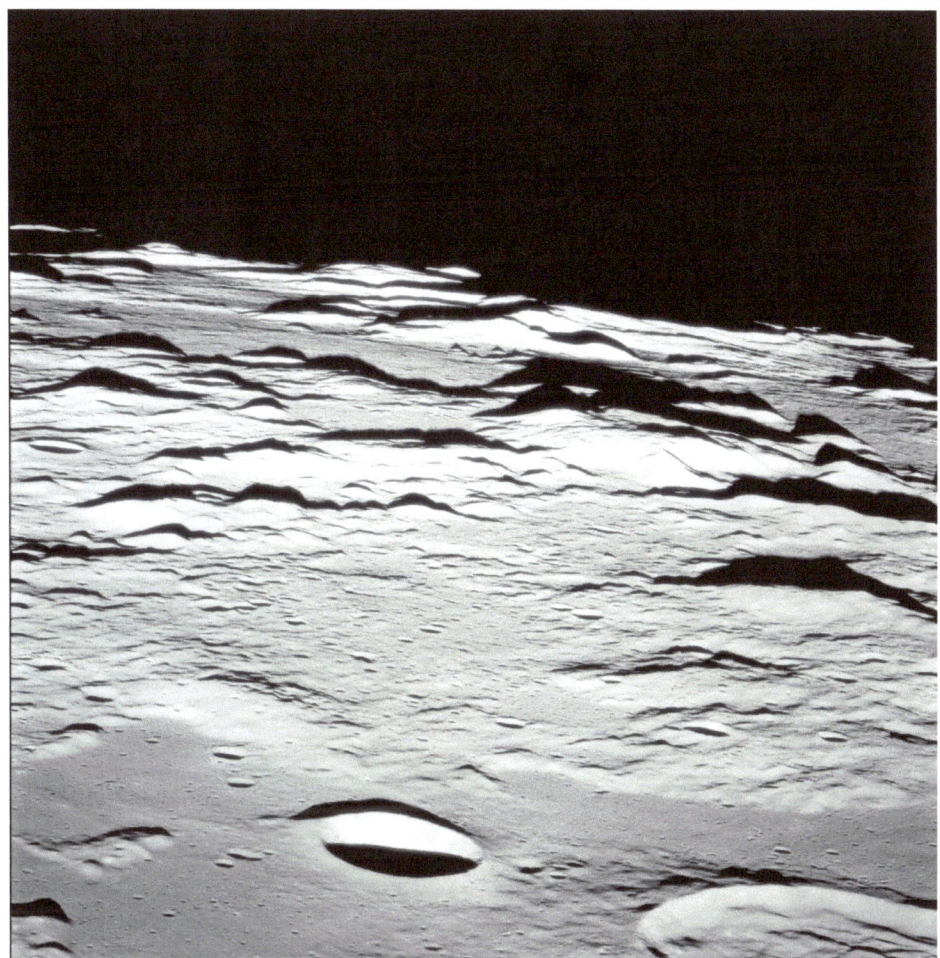

View of Moon, Sabine and Schmidt Craters. This image is part of a west looking oblique sequence of images taken from the Command and Service Module as it traveled at approximately 60 nautical miles (NM) orbital altitude above the Moon during the Apollo 11 mission. This sequence commences at 35 degrees East Longitude and continues to the nearside lunar terminator.

Apollo 11: 52nd Anniversary Pictorial

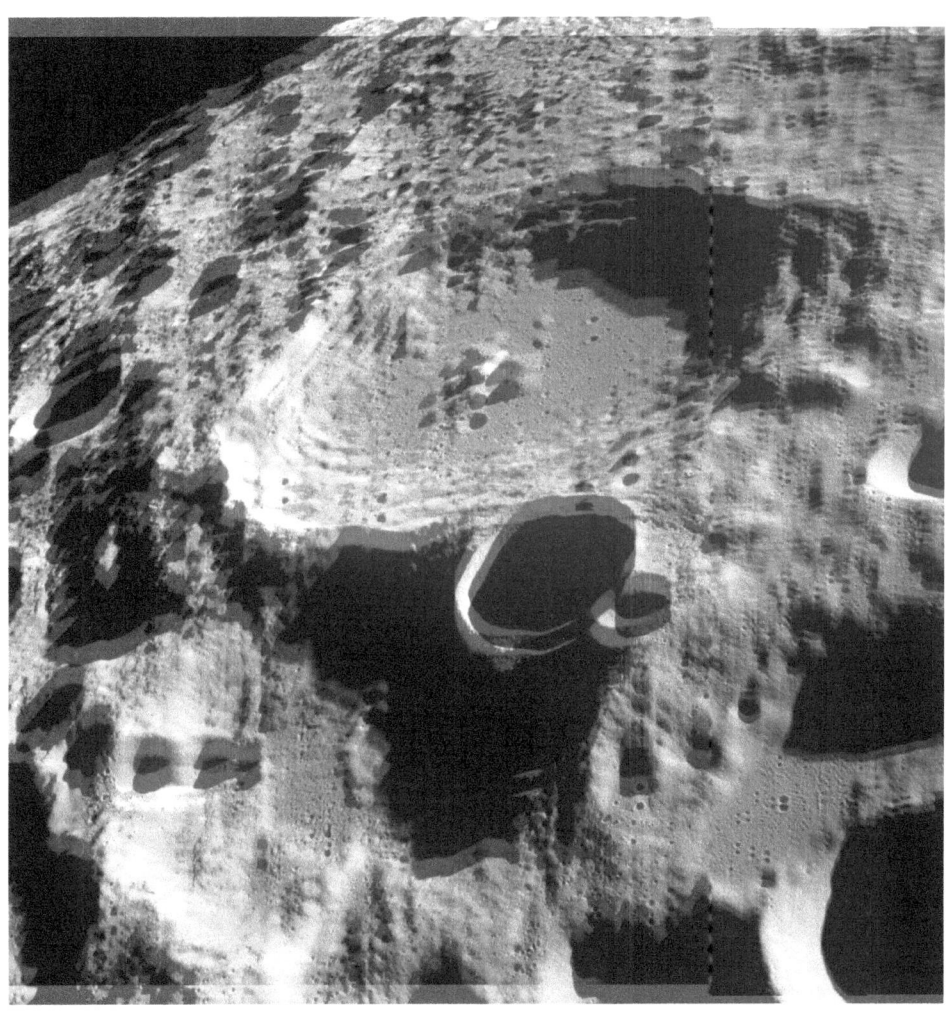

Lunar surface: Image taken at Tranquility Base during the Apollo 11 mission. Original film magazine was labeled S.

View of the Lunar Module ascent stage. The Earth is visible above the Lunar Module. Image taken at Tranquility Base during the Apollo 11 mission.

Apollo 11: 52nd Anniversary Pictorial

Close up view of Lunar Module ascent stage. Image taken at Tranquility Base during the Apollo 11 mission.

Lunar horizon from Tranquility Base, the Lunar Module landing site. Lunar Module thrusters in foreground. Unites States flag and television camera visible on lunar surface. Image taken from inside the Lunar Module during the Apollo 11 mission.

Apollo 11: 52nd Anniversary Pictorial

Lunar horizon from Tranquility Base, the Lunar Module landing site. Lunar Module shadow is visible on lunar surface. Image taken from inside the Lunar Module during the Apollo 11 mission.

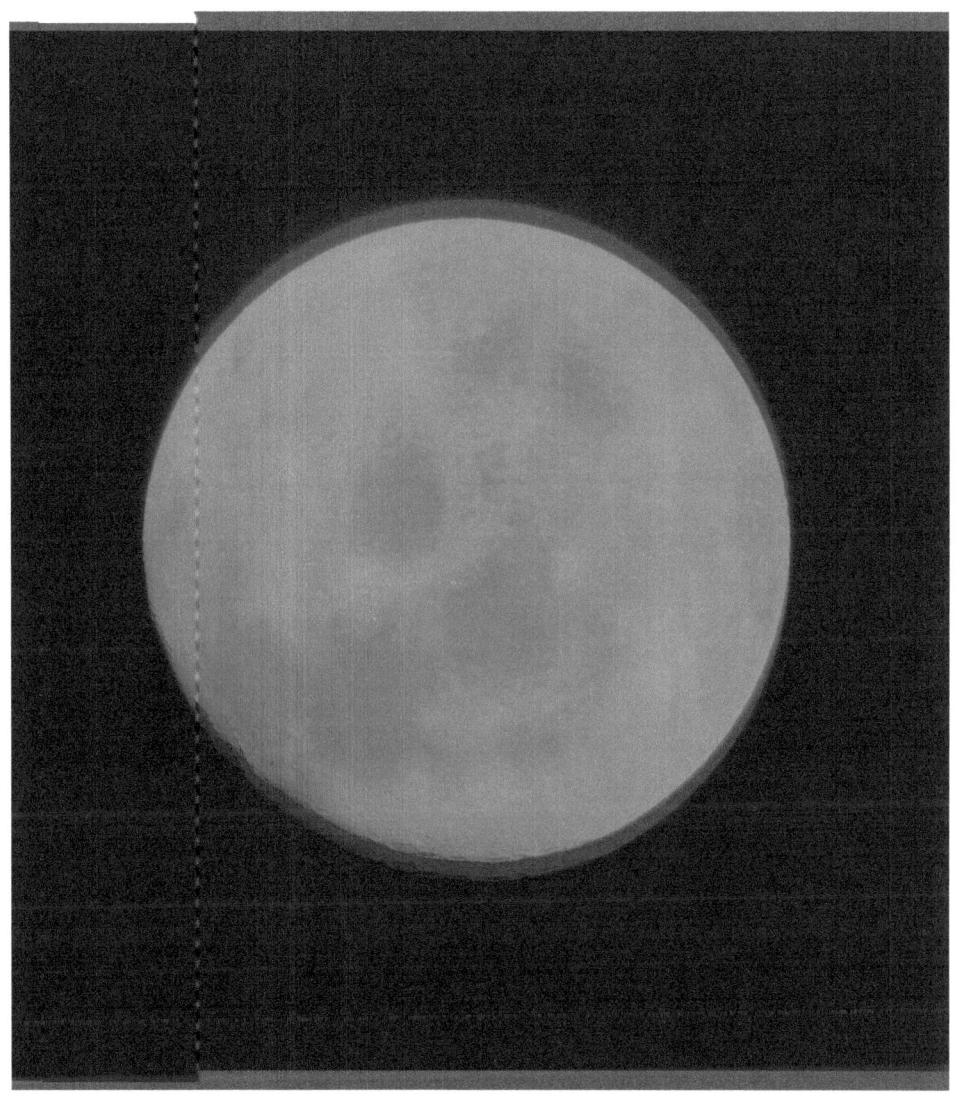

View of Moon: This image was taken from the Command and Service Module after Trans earth Insertion during the Apollo 11 mission.

Apollo 11: 52nd Anniversary Pictorial

View of Moon. This image was taken from the Command and Service Module after Trans earth Insertion during the Apollo 11 mission.

Donald Yates

View of the Earth terminator. One third of Earth sphere illuminated, East Africa visible. Image was taken after the trans earth insertion as the Apollo 11 crew traveled back to Earth.

Apollo 11: 52nd Anniversary Pictorial

Recovery

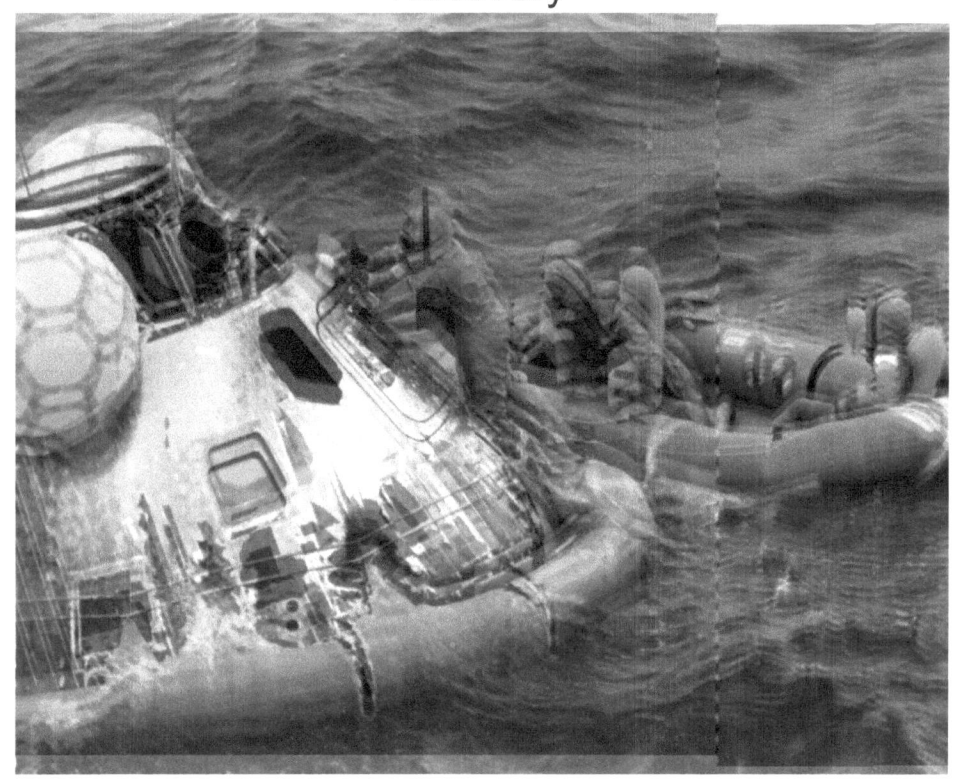

Lt. Clancy Hatleberg of Underwater Demolition Team Eleven decontaminates the exterior of the Apollo 11 command module after the astronauts Neil Armstrong, Michael Collins, and Edwin Aldrin [Buzz Aldrin] left the module and went aboard the inflatable board. Lt. Hatleberg will also decontaminate the astronauts.

Members of the Navy's Underwater Demolition Team Eleven work to recover the Apollo's crew and reentry capsule.

Apollo 11: 52nd Anniversary Pictorial

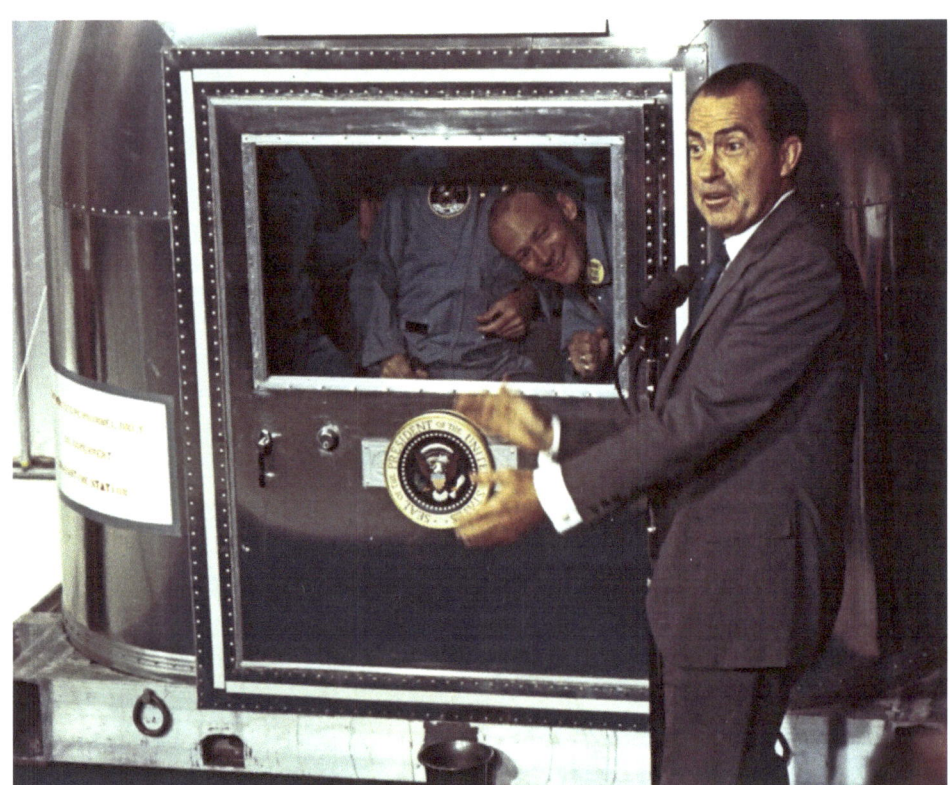

NASA/APOLLO. ABOARD THE USS HORNET - President Richard M. Nixon appears to be holding the Presidential Seal which is attached to the door of the Mobile Quarantine Facility during his greeting today to Apollo 11 astronauts, left to right, Neil A. Armstrong, Michael Collins, shoulders only, and Edwin E. Aldrin, Jr.

Donald Yates

Appendix A Star Chart

The following page displays the star chart for any course correction needs on the historic first moon landing by Apollo 11 in July of 1969.

Apollo 11: 52nd Anniversary Pictorial

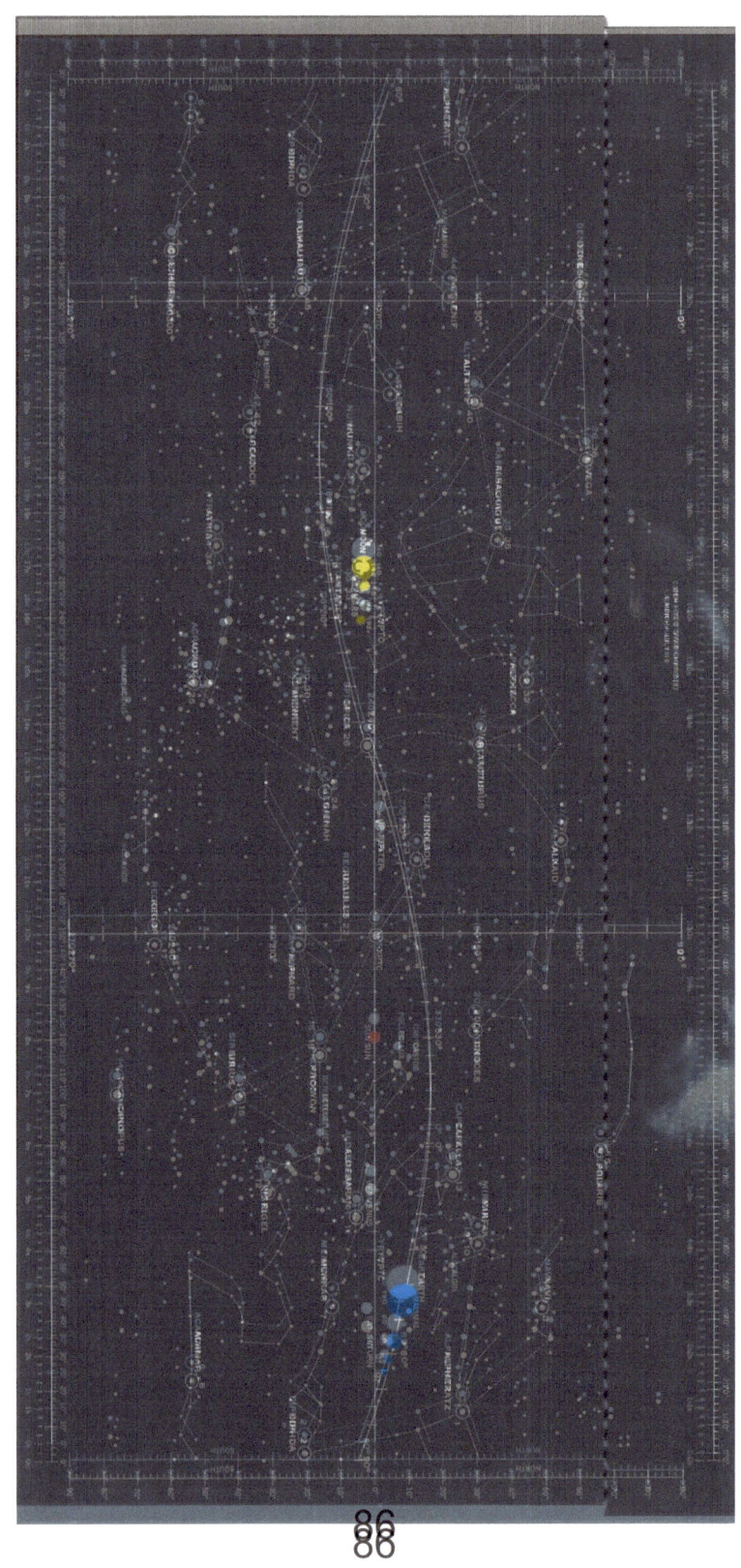

Appendix B Flight Profile

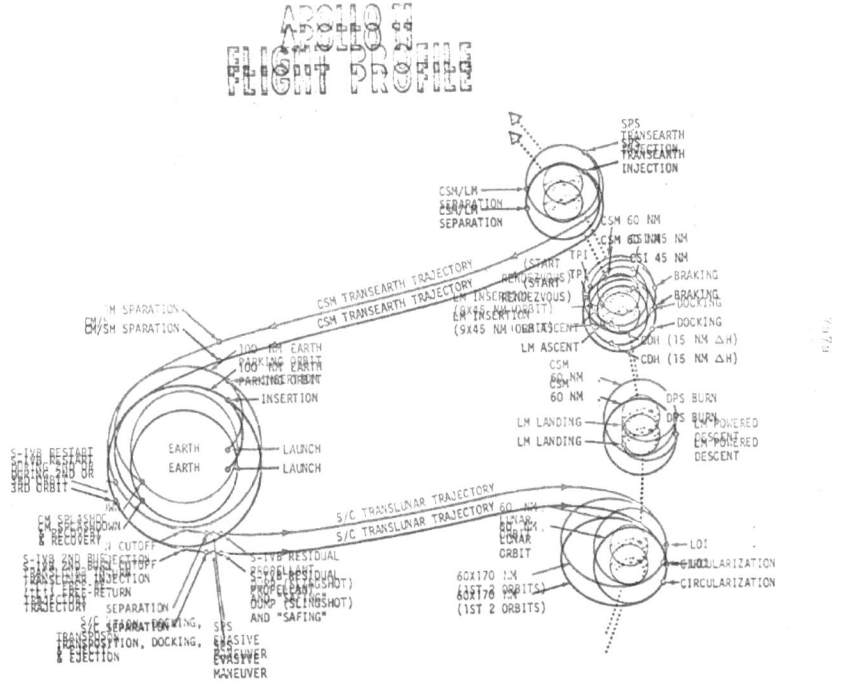

This is a diagram of Apollo 11 flight trajectory in relationship to the Moon's location:

Apollo 11: 52nd Anniversary Pictorial

Appendix C Flight Plan

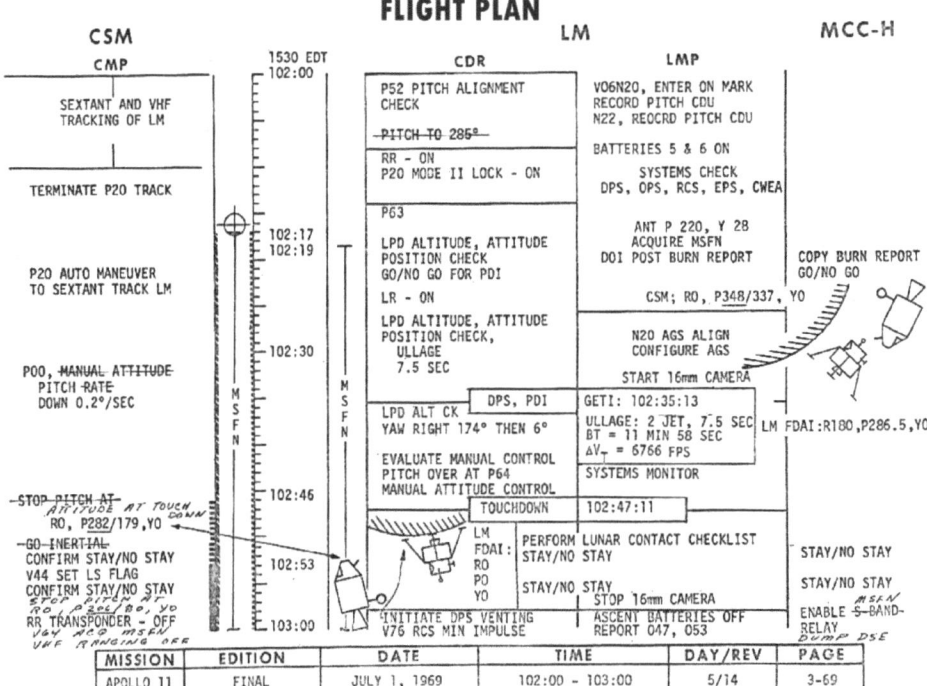

Apollo 11 flight plan dated 1 July, 1969. Go , no go decisions covering the Command Service Module (Command and Service Module) and the lunar Module . Page 3 of 69.

Donald Yates

References

Gilligan, M. 10 of the Most Watched Television Events in History. https://didyouknowfacts.com/10-of-the-most-watched-television-events-in-history/

Marko, P. Apollo 11 Restored EVA Part 1 [Video] https://vimeo.com/14275570

Marko, P. Apollo 11 Saturn V Launch (HD) Camera E-8 [Video]. https://vimeo.com/4366695

Apollo 11: 52nd Anniversary Pictorial

Other Books by the Author

Thank you for reading *Apollo 11: 52nd Anniversary Pictorial*.

Titles of other history books I have written are listed below with the Amazon Standard Identification Number:

The Life of the M1 Abrams Tank
ASIN: B098TZXTZ6

Marine Corps Tarawa Operations: A Photo Gallery
ASIN: B095Z45HSZ

Marine Corps LAV-25 Operations: A Photo Gallery
ASIN: B095NGQL6D

Guns of Gore: U.S. Field Artillery Howitzers of the 20th Century
ASIN: B093TQLVHB

Marine Corps Helicopter Assault: Vietnam: A Photo Gallery
ASIN: B095LMVC3F

Marine Corps Tanks in Vietnam: A Photo Gallery

Donald Yates

ASIN: B0951YBZ1L

UNCOMMON BRUTALITY: 101 Incidents of Japanese War Crimes During World War II Volume 1
ASIN: B08W1TM5TQ

UNCOMMON BRUTALITY: 101 Incidents of Japanese War Crimes During World War II Volume 2
ASIN: B08WLC8GY1

UNCOMMON BRUTALITY: 101 Incidents of Japanese War Crimes During World War II Volume 3
ASIN: B092PTHVXR

FROM BOSTON TO TOKYO: The History of the USS San Diego
ASIN: B08VBPMNX4

You may reach Donald Yates at yatespub@gmail.com.

www.ingramcontent.com/pod-product-compliance
Lightning Source LLC
Chambersburg PA
CBHW051914210526
45473CB00006B/2010